Tupak Ernesto Obando Rivera

Introduction to Geobiology

Tupak Ernesto Obando Rivera

Introduction to Geobiology

ScienciaScripts

Imprint

Any brand names and product names mentioned in this book are subject to trademark, brand or patent protection and are trademarks or registered trademarks of their respective holders. The use of brand names, product names, common names, trade names, product descriptions etc. even without a particular marking in this work is in no way to be construed to mean that such names may be regarded as unrestricted in respect of trademark and brand protection legislation and could thus be used by anyone.

Cover image: www.ingimage.com

This book is a translation from the original published under ISBN 978-613-9-18908-3.

Publisher:
Sciencia Scripts
is a trademark of
Dodo Books Indian Ocean Ltd. and OmniScriptum S.R.L publishing group

120 High Road, East Finchley, London, N2 9ED, United Kingdom
Str. Armeneasca 28/1, office 1, Chisinau MD-2012, Republic of Moldova, Europe
Printed at: see last page
ISBN: 978-620-6-30282-7

Chapter 1 Geobiology

Rio Tinto (Huelva, Spain), a privileged site of research into major problems geobiological.

Geobiology is an interdisciplinary scientific field that explores the interactions between life and the Earth's physico-chemical environment. It can also be defined as an interdisciplinary study between life sciences and earth sciences. It has important similarities with biogeology, but the latter would have a narrower scope.[1]

Researchers involved in geobiology belong to fields such as geochemistry and biogeochemistry, mineralogy, sedimentology, climatology and oceanography, soil science, palaeontology, microbiology, physiology and genetics, ecology and in general to all those geological specialities in which it is important to understand the influence of living beings, and those biological or environmental specialities in which the physical environment is involved.

History

Geobiology is admittedly a recent discipline, but the elements of its justification are already to be found in the foundational work of modern geology, James Hutton's *Theory of the Earth* (1788). The term was coined by L. Baas Becking (1934), who presented the problems of the field in much the same way as they are conceived today. Shortly before, Vladimir Vernadsky had given the term biosphere its current meaning, pointing to the integrated character of geological and biological processes on Earth. The Gaia hypothesis, proposed by J. Lovelock in 1969, although controversial in some respects, contributed to the expansion of interest in this frontier field.

Fields of study

At the interface or boundary between the life and earth sciences, geobiology coincides with the environmental sciences, specifically focusing on the influence of living beings on the earth system, and on the constraints that the planet's physical processes place on the evolutionary and ecological development of life. An exemplary case of a geobiological problem is that of the evolution of the composition of the Earth's atmosphere as the types of metabolism evolve, for example with its conversion into an oxidising atmosphere under the influence of the oxygenic photosynthesis "invented" by cyanobacteria.

Biogeochemical cycles

A field of study that should be integrated as an important part of Geobiology is biogeochemical cycles: these are cyclical movements of the elements that form biological organisms and the geological environment and involve chemical change.

Biogeology

Biogeology is the science that studies the interactions between the terrestrial biosphere and the lithosphere.[1]

Pyrite

It can also be defined as an interdisciplinary study between biology and geology. It has important similarities with geobiology, but the latter would have a broader scope.[2]

Biogeology examines biotic, hydrological, and earth system components, in their mutual relationship, to help understand the Earth's climate, oceans, and other effects on geological systems.[3]

For example, bacteria are responsible for the formation of some minerals such as pyrite, and can act to concentrate economically important metals such as tin and uranium. Bacteria are also responsible for the chemical composition of the atmosphere, which affects the weathering rates of rocks.

Before the Late Devonian period, plants other than lichens and bryophytes were rare. At this time, large vascular plants evolved, growing up to 30 metres tall. These large plants changed the atmosphere, and altered the composition of the soil by increasing the amount of organic carbon. This helped prevent the soil from being washed away by erosion.

One of the most famous biogeologists in the United States was Dr. Preston Cloud, a professor at the University of California at Santa Barbara. Cloud received a NASA research grant to examine moon rocks brought back by the Apollo missions.[4]

Some pseudo-sciences such as dowsing try to include their contributions within Biogeology as they talk about interactions between energies or waves supposedly emitted by earthly materials that would be perceived or would end up affecting humans and other living beings.

Geobiology could be defined as the discipline that studies the interaction between the Earth and living things, or rather, the effect that each area of the Earth has on the beings that inhabit it. Geobiology does not belong to any strictly academic branch of science, nor does it claim to, although it does draw on key concepts from geology and biology, as well as physics and neuroscience. In fact, the first to investigate this in Europe came from a medical background: the Englishman Havilland, the German Gustav Freiherr Van Pohl, Dr. Hager, president of the Austrian Scientific Association of Doctors of Medicine, or the German physician Ernst Hartmann, among others. At the beginning of the 20th century, many people began to rediscover the relationship between staying in geophysically altered places and various disorders and diseases.

Geobiology deals in particular with everything that can affect the health or well-being of people in their environment and, more specifically, in their habitat. Its name comes from the union of the words Geo (Earth) and Bio (Life). For this reason, it is also known as the science of habitat.

For centuries, man has been able to understand the influence of the Earth's magnetism on our health, detect geopathogenic zones and decide which is the healthiest place to live. He has been aware of the natural radiations emanating from the ground and their effects on health, and has been able to detect them by biosensing methods (which has been used since the dawn of time to search for water and metals in all cultures of the world, not only in the West, using dowsing rods, dowsers and dowsing).

Nowadays, geoenvironmental health takes all this ancestral knowledge, integrates it with current scientific knowledge and materialises it in concrete protocols to holistically address the influence exerted on us by our environment and identify the factors and parameters that are dangerous to our health. With the help of geomagnetometers, detectors of electric and magnetic fields and variations in natural radioactivity, it is possible to determine in a contrasted way the best location for a person in their habitat (home or office) and thus avoid placing them in areas of harmful influence that could cause harmful effects on their health.

Geomagnetic networks and geophysical disturbances

The natural factors that have traditionally been studied from Geobiology and that are nowadays studied with modern equipment from Geoenvironmental Health are:

- Geophysical disturbances: geological faults and fractures in the ground, contact areas between different types of materials, groundwater flows and other subsurface features can cause both local electromagnetic disturbances in the vertical of the subsurface, and local electromagnetic disturbances in the vertical of the subsurface. phenomena, such as changes in ambient radiation levels.
- Hartmann lines: natural geomagnetic network whose lines of force form a north-south oriented grid with cells of approximately 2 by 2.5

metres.

- Curry Lines: a natural geomagnetic network whose lines of force are oriented northeast-southeast and southeast-northwest, approximately every 6 to 8 metres.

- Environmental radioactivity from rocks and soil materials, which can often result in high concentrations of radon gas, a highly carcinogenic substance, according to the World Health Organisation.

Geophysical disturbances

The subsoil on which our homes and offices are built, or on which we plan to build a property, can affect the electromagnetic environment on the surface. Beneath the ground we walk on, there may be faults, joints or cracks; there may be different types of materials in contact with each other, causing physical and chemical reactions that rise to the surface; there may be underground streams, aquifers or bodies of water, so that the geophysical fields in the environment may vary.

In short, there are a number of geophysical factors that can influence our environment. Since our vital organs function through electromagnetic mechanisms, electromagnetic variations in our daily environment interfere with our vital rhythms and can weaken our health, opening the door to disease.

Geological faults

The earth's crust is in continuous movement due to seismic and tectonic forces. These forces produce faults, fissures, cracks, diaclases.... These are discontinuities or fractures in the rocks of the subsoil, and these alterations can be present anywhere beneath the ground in which we live; when this happens, the parts of the ground that have been fractured bring surfaces of different natures into contact with each other; they often even form underground cavities. In the vertical of these phenomena, due to the law of least resistance, a whole set of energies emanate from the subsoil, strong gamma radiation and even radioactive gases. This has ionising effects on the surface atmosphere, and also influences the magnetic field of our surroundings, causing variations of varying magnitude.Groundwater

Groundwater represents a significant fraction of the total mass of water present on the continents of our planet. Water currents underground, aquifers, water pockets, sinkholes and seepages fill underground cavities and circulate in underground galleries, but also occupy underground pores and crevices. Their presence beneath the ground we walk on decreases the value of the earth's magnetic field and increases gamma radiation (radioactivity), as well as causing strong variations in the ionisation of the air. Their area of influence depends on the size of their flow: the larger the flow, the larger the area affected on the surface.

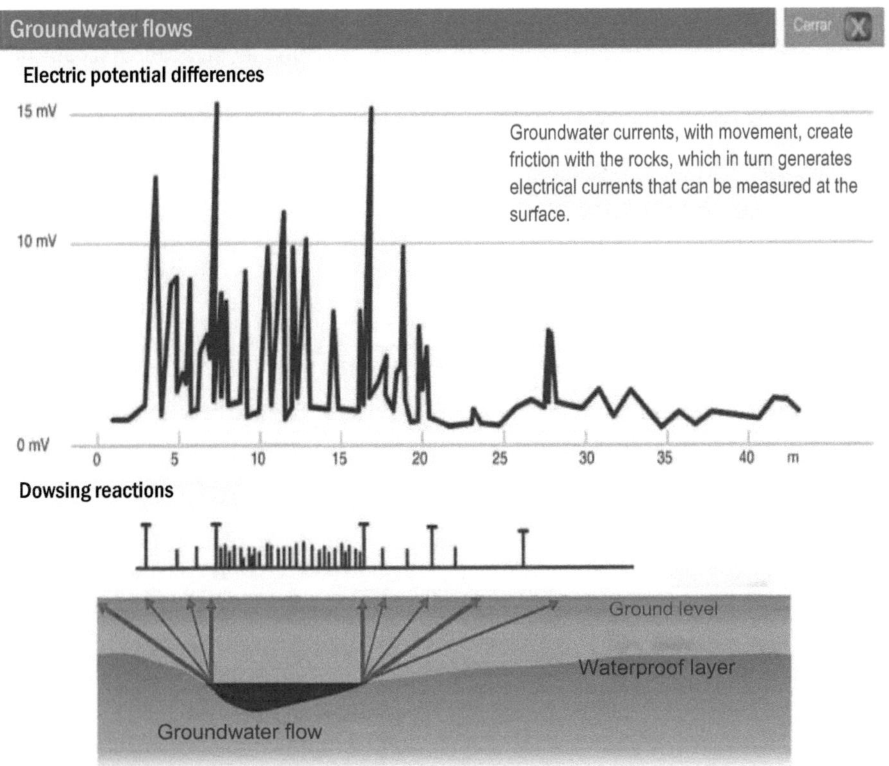

Given the dynamic behaviour of water, it is difficult to predict whether a

subsurface stream or aquifer will remain in place for a long time or whether it

will percolate and disperse. In any case, it is of less interest to know the nature of the aquifer but rather the effects it has on the surface.

We are living in, as it is the surface variables that affect our health, not the presence of water per se.

Of course, if several of these geopathogenic factors are combined, such as a groundwater flow, a Hartmann line and a Curry crossing, the harmful effect of the area will be much greater. A geoenvironmental health expert can detect all these risk factors and design solutions that preserve our well-being.

Environmental radioactivity and radon gas

Radioactivity is a natural physical phenomenon that occurs when the atomic structure of any substance does not have the correct balance between protons and neutrons. The French physicist Henri Becquerel discovered this phenomenon in 1896 when he noticed that certain uranium salts emitted radiation spontaneously, which veiled photographic plates wrapped in black paper. Later, the Curies discovered other radioactive substances such as thorium, polonium and radium. Marie Curie was awarded the 1903 Nobel Prize in Physics, the first woman to win the prize, for her studies on radioactivity.

We tend to identify the phenomenon of radioactivity with certain man-made installations, such as nuclear power plants or radioactive environment. Natural radioactivity reaches us from the sky (cosmic radiation), from the air we breathe (which contains carbon and may contain radon gas) and from the ground (where uranium and thorium may be present). Our bodies also contain radioactive elements: for example, we need potassium to survive and we get it from common salt.

However, natural radioactivity becomes a risk to our health when it increases to a degree that our bodies are not prepared to assimilate. High levels of radioactivity are often found in our everyday environment due to the mineral composition of the subsoil in localised areas. The ground we walk on or on which our homes are built may contain granite, clays, etc., which have a high concentration of uranium which, as we have already seen, is highly radioactive. This mineral is also present in certain construction and decoration materials, such as some types of stoneware or ceramics, or certain types of cement.

In its natural decay process, uranium emits radon gas, which is officially classified by the World Health Organisation as the second leading cause of lung cancer in the world. The result is that our home or office can register high levels of naturally occurring radioactivity.

and radon gas without us being aware of it, since radon gas is odourless, tasteless and invisible. A high concentration of radon gas in the air, which is

we breathe in saturates our lungs with carcinogenic radioactive elements.

But what exactly is Geobiology?

It is still a little-known word and is often confused with geology or biology. And, in fact, it has a bit of both.

Geobiology is an interdisciplinary science (although some classify it more as an art or technique) that studies the relationship between 'gea' -earth- (the energies coming from the earth) and 'bios' -life- (the living beings that inhabit it).

In other words, geobiology is concerned with studying the relationship between the **quality of living space and the health of the people who inhabit or frequent that space.**

Translated into practice, this means that geobiologists study electromagnetic pollution, toxic materials used in construction, the effects of terrestrial radioactivity on houses, human-generated artificial radiation and how they all affect us.

This is because where we live and work directly affects our personal vitality.

Where we live and work directly affects our vitality.

staff.

LIVING SPACE

Your living space, whether you are aware of it or not, is influenced by:

- **Natural objects**: earth, water, plants, animals...

- **Artificial or man-made objects**: houses, furniture, vehicles.

- **Unseen influences**.

As for these **non-visible influences**, some of them are already **known and measurable today**, such as weather, chromatic radiation, radioactive radiation, sound vibration and electromagnetism.

However, there are still other **non-visible influences** which, although they have been studied for years, are still **considered unknown** and therefore difficult to disseminate and to develop measuring devices for. These are: geobiological networks, telluric discontinuities, shape emissions.

There are various types of geobiological disturbances that can affect you, but the ones that have the most consensus in the sector are:

- Hartmann geomagnetic network (also called global network).

- Red Curry (or diagonal).

- Groundwater circulation.

- Cracks, fissures, faults in the subsoil.

- Buried water mains or power lines.

- Cosmotelluric chimneys

-

You can read a brief description of each in the post "What is a geopathy".

Geomagnetic grids and especially **crossings**, influence our organism depending on the more or less prolonged stay at these points. Therefore, it is very necessary to **prospect those areas where you spend many hours, such as your bedroom and your workplace.**

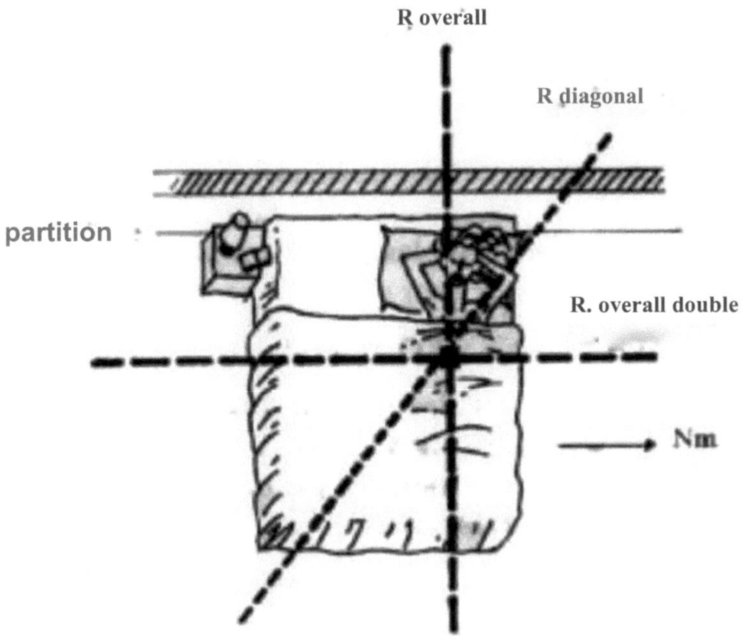

In a case like the one in the picture, the person could be affected with:

- Migraines

- Heart problems

- Insomnia

- Fatigue

- Nervousness

- Cancer

But in general, it would enhance any weaknesses that their

organism.

How to do a Geobiological survey of a house

A complete geobiological survey will analyse, measure and solve the sources of electromagnetic pollution, both natural and man-made, in your home.

A complete geobiological survey will analyse, measure and solve the sources of electromagnetic pollution, both natural and man-made, in your home. Click to tweet First of all, the geobiologist will walk around your house with rods and use his personal sensitivity to first find the global grid (Hartmann) and the diagonal grid (Curry). He will mark them on the scale plan of your house. Under this basic scheme, he will then look for cosmotelluric chimneys, faults and underground watercourses. A plan of a geobiological survey can look like this way:

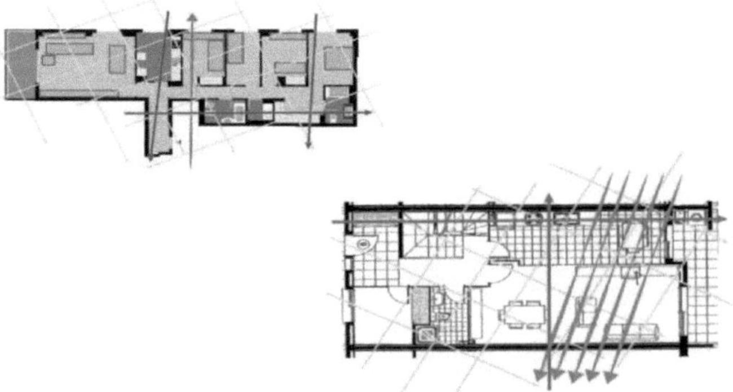

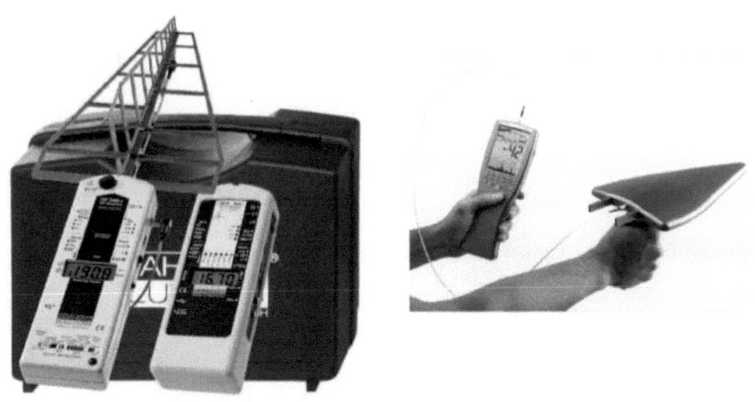

Thanks to this study, which **analyses natural radiation**, it is possible to determine which areas of your home are neutral and where you can place your bed, study area and sofa without danger to your health. In other words, **the places where you spend the most time**. And you can locate the **geopathogenic areas** of your home to take them into account and avoid them when redesigning the interior of your home.

Full study

A complete study will then follow with the **search for, measurement and analysis of artificial radiations**. What I mentioned before as non-visible, but known influences).

For this purpose, the geobiologist has **professional measuring equipment** with which to analyse **high frequency electromagnetic pollution** (telephone antennas, radars, mobile phones, wifi, bluethooth, cordless phones, etc.), **low frequency electrical pollution** (the electrical installation in your home, sockets, cables, lamps, household appliances, low and medium voltage lines outside the home, etc.) and **low frequency magnetic pollution** (high, medium and low voltage lines, transformers, etc.).

After these studies, **personalised solutions** will be determined **for each case**. **Biocompatibility will be** sought at all times. In other words, the positive influence on the health of the people who live in or frequent that place.

Geobiology is a science that gathers the deepest knowledge about the relationship between living beings and the different energies and vibrations that make up life.

Cosmotelluric radiations show us the interrelation of all natural phenomena and biological processes. The human being does not depend only on the food he eats or the air he breathes, but there is a subtle, constantly vibrating energy in the universe that structures and animates all living beings and the earth itself.

Throughout the centuries, man has been aware of this and the Romans already carefully chose the location of their cities. The Egyptians, great connoisseurs of telluric radiations, ensured that their constructions were oriented according to them, as did all the great civilisations.

Recent work has shown that we constantly receive vibrational waves from the centre of our planet, which have been detected by means of sensitive detectors (galvanometers). The electrical resistance of the individual's skin has been measured, showing that it varies depending on the location.

Geobiology is the science that studies the effects of telluric radiation, water, gas or air currents, faults, magnetic lines, electromagnetic disturbances, electrical pollution and other...

It has been studied by physicists, biologists and doctors. Postgraduate courses have been given at universities (Ramón Llull, UPC, Barcelona, Rovira y Virgili and others) for architects, engineers and health personnel, and more and more professionals are using this knowledge to design or build a house.

HISTORY

The ancient Chinese forbade the construction of dwellings for people or shelters for animals on what they called dragon veins.

The Romans used to graze flocks of sheep for long periods of time on land where they intended to found a city. After the animals had been slaughtered, they studied their livers, the condition of which gave them information about the quality of the land.

We know that the North American Indians let their horses graze freely and watched their favourite places carefully. (This information is not valid today as the animals are exposed to more stimuli as a result of modern city life).

Today, the Tuareg, a nomadic desert tribe, still use the observation of chosen sites by their dogs.

Gustav von Pohl after years of study states in his book: Erdstrahlen als krankheitserreger-Forschungen auf Neuland, Munich 1932: "the cancer cases in Vilsbiburg occurred in dwellings exposed to electronegative radiation emanating from groundwater streams".

Pierre Cody, a scientific engineer and pioneer in the investigation of telluric radiation, focused his research on the study and analysis of the ionisation of air in the vertical of underground water veins. In 1935 he was the first to point out the rhodon momo gas as the cause of lung cancer.

Rémi Alexander, French architect and geobiologist, after personally experimenting with the effects of different radiations and deepening previous research on the subject, undertook work on the application and dissemination of geobiology and bio-construction. The results of his work have been published in two works: Votre lit est-il a la bonne place? and Votre maison.

Kathe Bachler, as a pedagogue and secondary school teacher in Austria, focused part of her work on the impact of radiation on school performance, as well as on the disorders of teachers and pupils, and on the study of pathologies caused by groundwater currents, hartmann and curry.

Ernst Hartmann, a physician at the University of Heildelberg, Germany, was the discoverer of the Hartmann network. In his work: krankheit als Standort problem, he describes some of the most classic cases in the geobiological literature.

Association for Geobiological Studies GEA

Centro Mediterráneo de Investigación Geobiológica, directed by Mariano Bueno, expert in geobiology, ecobioconstruction and ecological agriculture. He is the author of the works: vivir en casa sana and el gran libro de la casa Sana (Martínez-Roca publishing house).

Geonatura: Association of geobiologists, architects... dedicated to the research and teaching of geobiology, directed by Pere Vila, a geobiologist with more than 30 years of experience, who together with his team and doctors and therapists from all over Spain work on the research and statistics of geobiology and its relationship with chronic and degenerative pathologies.

RADIESTHESIA

It is the modern name for an ancient art such as the world art of the dowsers.

If we analyse the word dowsing, we find it etymologically formed by the Latin root (Radius), which means radiation, and the Greek root (stesia), which means sensitivity.

The rod and pendulum have always been essential weapons of the dowser to detect geopathic zones, through the education of his brain during years of experience to detect this type of radiation and to produce a physical stress reaction which is reflected by the movement of the pendulum or the rods.

Natural radiation can only be detected by personal perception, while artificial radiation is detected by machines, which detect

THE ASSAULT

In today's world, our bodies are subjected to a variety of stresses:

- Poor nutrition

- Stress

- Emotional aggression.

- Tobacco

- Environmental Pollution

- Electromagnetic radiation

- Geopathies

- Other

Statistical studies indicate that 85% of clinical cases of chronic and degenerative diseases correspond to people who have been exposed to some

kind of intense geopathy for a prolonged period of time.

Chapter 3 WHAT ARE GEOPATHIES?

We speak of geopathy when a person's organism receives an excess of radiation from the subsoil and/or the environment during a certain period of exposure.

These factors influence to a greater or lesser degree depending on the nature of the individual, the strength of the radiation and the time of exposure.

TYPES OF GEOPATHIES

. UNDERGROUND WATER FLOWS:

Today's geology recognises the circulation of water in karst terrains through fissures in terrains with other geological structures. These are known as water veins, which are the source of water supply to the wells.

They are present and are of great importance in more than 80% of all studies carried out on chronic and degenerative diseases. Their radiation affects the defences and as a consequence of the alteration of the regulation system that the individual possesses, it favours the appearance of various dysfunctions and behind them the

These are the ones that appear in red on the map, and we must avoid our exposure especially in places where we spend a long time: bed, desk, sofa...

.. HARTMANN NETWORK :

This grid is oriented according to the earth's magnetic field, in a north-south and east-west direction with a spacing that varies from 1.60 m. to 2.60 m. It has a thickness of about 20 cm. It passes through everything: land, buildings, plants, animals and people, and can be found everywhere.

Being exposed to a single Hartmann's line does not affect health because of its weak intensity, it is only pathogenic when it is a crossbreed that affects and with several hours a day of exposure to it, in vital parts of the body (head and trunk).

The dimensions and activity of Hartmann lines are altered by atmospheric changes, seismic movements and other noxious wave emissions and by the normal displacement of the Earth's magnetic north, in the same way as curry lines. They are therefore considered to be less harmful.

These are the ones shown in blue on the map.

- CURRY LINES:

They form a grid of approximately 4 to 8 m in a NE-SW and NW-SE direction, with a thickness of about 40 cm. Their pathogenic effects are greater than those of the Hartmann grid. They are not caused by the ground and are electromagnetic in nature.

Maximum action occurs when there is additional crosstalk with radiation from the earth, or with each other.

These are the ones shown in purple on the map.

- ARTIFICIAL ELECTRICAL AND ELECTROMAGNETIC RADIATION:

Electromagnetic fields produced by external radiation: power lines and transformers, radiofrequency waves from mobile phone, radio and television antennas.

Radiation inside the house: cordless phones, radio alarm clocks, wireless networks, wifi and small transformers of electrical appliances.

The radiation produced by outdoor mobile phone antennas can be shielded by 80% of its intensity, by placing a metal grid on the wall where the radiation comes from, or by painting it with a special metallic paint, and if the radiation enters through the window we can place a special curtain, made with an invisible metallic thread.

SYMPTOMS

The diagnosis of a geopathy, taking into account the symptoms, is very difficult, because of the large number of symptoms that a person may have in a geopathogenic zone. It affects everyone equally, but it will depend on the nature, the physical condition of each person, whether they have been exposed to a geopathy before, even at the time of their gestation, and the most sensitive hereditary part of each person to manifest some or other symptoms with more or less intensity and severity.

SLIGHT

- nervousness
- irritability

- depressed mood

- difficulty falling asleep

- insomnia,

- nightmares

- muscle whiplash when falling asleep

- pollakiuria in the middle of sleep

- premature awakening

- tiredness on waking

- chronic fatigue

- cold feet, cold sensation in the back

- paraesthesia in hands and feet, cramps

- palpitations

- gnashing of teeth

- chronic rhinitis...

MODERATE:

- tenacious insomnia

- chronic tonsillitis

- bronchitis

- asthma

- sweating for no reason

- tension in the back of the neck

- back and lumbar pain

- lumbago for no apparent reason

- chest pain

- rheumatic pains

- dyspnoea, headaches

- migraines

- dizziness

- neuralgia

- demineralisation osteoarthritis

- asthenia all day long

- deep depression

- latent aggression

- suffer from many phobias

SERIOUS:

- circulatory and cardiac problems

- impairment of the defence system due to difficulty in eliminating toxins

- exhaustion

- nervous

- suicidal thoughts

- personality splitting

- schizophrenia

- panic attack

- multiple sclerosis

- autoimmune diseases

- degenerative diseases

- cancer

Health. Marco Vitrubio proposed in his treatise "De Architectura", that architecture rests on three basic principles, the Venustas (beauty), the Firmitas (firmness) and the Utilitas (utility), being architecture, a balance between these three variables and the absence of one of them, would mean that such a work could not be considered as such. For us, if there is no health within a space, it ceases to make sense, neither for its utility, nor as beauty. We believe that La Salutis (health) must be the fourth basic principle and be on the same level as the other principles, since without health it is not possible to feel the emotions that build the experience of a place.

Geobiology. This is the science that studies the relationship between the earth and living beings. It studies the influences of the energies emanating from the earth, those coming from cosmic radiation and the energies generated by human activity and the biological processes inherent to human beings, focusing mainly on the analysis of the energetic and vital quality of inhabited space. Geobiology covers a wide field of subjects and draws its multidisciplinary knowledge from sciences such as Astrophysics, Geophysics, Hydrology, Biology, Electronics, Medicine, Architecture... combining ancestral knowledge of traditional wisdom with the most profound and recent knowledge, the result of scientific research, on the relationship between living beings and the different energies and radiations that make up life.

We ensure that the spaces we design are healthy for people by always carrying out geobiological studies of the health of pre-existing spaces, as they may have "invisible charges" such as problems of electromagnetic pollution from antennas or WIFI, electrical fields from the electrical system or lighting devices, subtle energies that alter rest and cause illnesses. We present some references where there is a large amount of information on all topics related to Geobiology, dowsing, ionising and non-ionising radiation and the health of habitable places.

Organización
Mundial de la Salud

WHO / World Health Organisation

The health effects of exposure to electromagnetic fields

Asociación de Estudios Geobiológicos is an independent non-profit association that focuses its work on research, analysis and dissemination of Geobiology as a science related to the health of people in relation to the habitat. It also organises introductory courses and training in Geobiology, informative conferences, technical training days, and meetings of members to promote exchange.

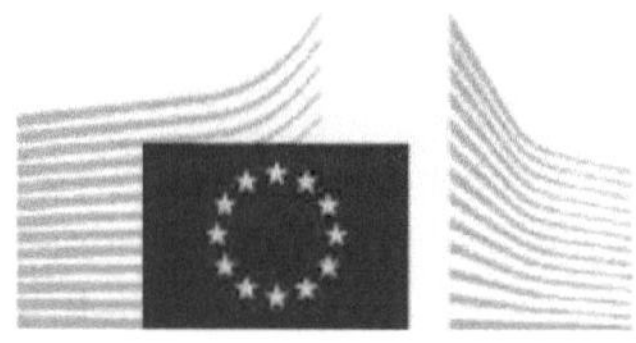

Science and Knowledge Committee of the European Union

Radon Prevention and Remediation

Private Foundation for Geoenvironmental Health

Foundation of the bioconstruction institute in Spain

TECHNICAL STANDARD OF MEASUREMENT IN BAUBIOLOGIE

SBM-2008

Independent, private initiative and non-profit organisation with the aim of

**Organización
Mundial de la Salud**

creating a healthier society.

WHO / World Health Organisation

Radon and its effects on health

Mariano Bueno

Expert in Ecological Agriculture, Geobiology, Bioconstruction and healthier living alternatives.Terra Aurea / Daniel Rubio

A space to disseminate information, experiences and serve as a link between

all

those interested.

dowsing and geobiology as a tool for

personal growth

and for study and research

of Mother Earth's energies as they interact with humans and the environment.

Universe.

Programme page on the different studies of Geobiology.

Programme on 2 October 2016 on channel Cuatro on the influences of

radiation on people's health in Spain.

Radon, A Naturally Occurring Radioactive Gas in Your Home
  Chair of Medical Physics. Department of Medical and Surgical Sciences.
University of Cantabria.

Map of radon exposure by municipality based on the predictive map

published by the Nuclear Safety Council.

Geobiology is the discipline (science) that studies the presence of natural geological disturbances (groundwater flows, faults, metal deposits, etc.) and telluric networks and points (cosmotelluric chimneys, Hartmann and Curry lines, etc.) and their impact on living beings, generally associated with the terrain on which buildings are erected (for example, how we are affected if a fault and a Hartmann line intersection converge under our bed).

Numerous studies and research have shown that prolonged exposure to these disturbances can be detrimental or even fatal to health. With the information provided by geobiology, we can seek solutions to live in spaces that are more biotically harmonious for their inhabitants.

The Earth is subjected to a constant bombardment of electromagnetic waves from outer space, most of which are absorbed by the atmosphere, but a significant portion of these waves manage to reach the Earth's surface, penetrating its interior.

Nowadays, due to technological progress, electric fields are sometimes greater than natural ones and are absorbed by the earth. These radiations, together with those produced in the interior of our planet and those that reach us from space, are sent back to the surface through what we know as telluric networks (mainly Hartmann and Curry lines).

Also, the very physical nature of the Earth's interior and surface, such as underground streams or water, faults and diaclases, cavities, metalliferous deposits, minerals such as granite, etc., generates a series of radiations and gases.

These radiations create a large quantity of positive ions in the vertical of the alteration, which further decompensate the ionic balance of the air we breathe. It must be taken into account that these radiations, both in quantity and composition, vary according to the type of geopathy and are very small quantities, almost undetectable, but enough to alter our health if we remain on them for a long time, mainly affecting the nervous, endocrine and hormonal systems. The cell acts as a resonant electronic circuit that is affected by natural and artificial electromagnetic frequencies. The highly positive ionised environment produced by electromagnetic radiation favours the appearance of free radicals that damage the cell membrane.

Furthermore, the vibratory environment generated in the vertical of the telluric disturbance is favourable for the propagation of pathogenic agents such as viruses, bacteria or fungi. Therefore, being exposed for months or years to certain telluric disturbances produces a double effect that favours the appearance of the disease: a favourable environment for the expansion of microorganisms and a reduced response capacity of our body.

Some of the **symptoms of being overexposed to one or more telluric disturbances are:**

- Difficulty in falling asleep .
- Difficulty in getting up .
- Waking up at the same time in the middle of the night.
- Lack of vitality .
- Feeling of lack of rest, even if you have slept.
- Same or similar symptoms in neighbours of the same letter of a building (vertical overlap).
- Illness or difficulty in recovering from illness. - Etc.

A star point, a cosmotelluric chimney, a water current or a fault, for example, could cause us to contract a disease if we are exposed to it for months or years, especially if it coincides with the place where we sleep. It must be taken into account that other factors such as emotional, energetic, genetic, viral, contaminating, intoxicating, traumatic, etc., contribute to the fact that the same geopathy does not affect all people in the same way.

With the vision and awareness that Geobiology brings us and with Healing as a tool we can contribute positively to change the energy of a house affected by geopathic disturbances and help to restore the health of its inhabitants. We can

carry out corrective measures, regardless of the location of the house,

References

- Discussion on geobiology, biogeology and geobiofacies. HongFu Yin, ShuCheng Xie, JianZhong Qing, JiaXin Yan and GenMing Luo. Science in China Series D: Earth Sciences. Volume 51, Number 11, 1516-1524, DOI:10.1007/s11430-008-0120-6, November 2006.

- Baas Becking LGM (1934) Geobiologie: of Inleiding tot de Milieukunde, Van Stockum and Zoon, Den Haag.

- Geobiology at Caltech. [1]

- Knoll, A.H. and Hayes, J.M., 1997, Geobiology: articulating a concept, in Lane, R.H. et al., eds., Paleontology in the 21st Century: Frankfurt, International Senckenberg Conference: Kleine Senckenberg, v.25, p. 105108.

- Knoll, A.H. and Hayes, J.M., 2000, Geobiology: Problems and Prospects, in Lane, R.H., Steininger, F.F., Kaesler, R.L., Ziegler, W., and Lipps, J., eds., Fossils and the Future: Paleontology in the 21st Century: Senckenberg-Buch n.74, p. 149.

- Kump, L., 2002, The Virtual Journal of Geobiology, from Elsevier Virtual Journal of

GeobiologyWebsite   :

Nealson, K., Ghiorse, W.A.,.2001, Geobiology: Exploring the Interface https://web.archive.org/web/20140227094534/http://earth.elsevier.com/ geobiology/.

- between the Biosphere and the Geosphere: American Society for Microbiology Report, 23 pp.

- Noffke, N., 2002, The Concept of Geobiological Studies: the Example of Bacterially Generated Structures in Physical Sedimentary Systems: Palaios, v.17, p. 531-532.

Table of contents

yes
I want morebooks!

Buy your books fast and straightforward online - at one of world's fastest growing online book stores! Environmentally sound due to Print-on-Demand technologies.

Buy your books online at
www.morebooks.shop

Kaufen Sie Ihre Bücher schnell und unkompliziert online – auf einer der am schnellsten wachsenden Buchhandelsplattformen weltweit! Dank Print-On-Demand umwelt- und ressourcenschonend produziert.

Bücher schneller online kaufen
www.morebooks.shop